~A BINGO BOOK~

Addition Bingo Book

COMPLETE BINGO GAME IN A BOOK

Written By Rebecca Stark

ISBN 978-0-87386-429-9

Educational Books 'n' Bingo

Printed in the U.S.A.

ADDITION BINGO DIRECTIONS

INCLUDED:

List of Terms

Templates for Additional Terms and Clues

2 Clues per Term

30 Unique Bingo Cards

Markers

1. **Either cut apart the book or make copies of ALL the sheets. You might want to make an extra copy of the clue sheets to use for introduction and review. Keep the sheets in an envelope for easy reuse.**

2. Cut apart the call cards with terms and clues.

3. Pass out one bingo card per student. There are enough for a class of 30.

4. Pass out markers. You may cut apart the markers included in this book or use any other small items of your choice.

5. Decide whether or not you will require the entire card to be filled. Requiring the entire card to be filled provides a better review. However, if you have a short time to fill, you may prefer to have them do the just the border or some other format. Tell the class before you begin what is required.

6. There are 50 terms. Read the list before you begin. If there are any terms that have not been covered in class, you may want to read to the students the term and clues before you begin.

7. There is a blank space in the middle of each card. You can instruct the students to use it as a free space or you can write in answers to cover terms not included. Of course, in this case you would create your own clues. (Templates provided.)

8. Shuffle the cards and place them in a pile. Two or three clues are provided for each term. If you plan to play the game with the same group more than once, you might want to choose a different clue for each game. If not, you may choose to use more than one clue.

9. Be sure to keep the cards you have used for the present game in a separate pile. When a student calls, "Bingo," he or she will have to verify that the correct answers are on his or her card AND that the markers were placed in response to the proper questions. Pull out the cards that are on the student's card keeping them in the order they were used in the game. Read each clue as it was given and ask the student to identify the correct answer from his or her card.

10. If the student has the correct answers on the card AND has shown that they were marked in response to the *correct questions,* then that student is the winner and the game is over. If the student does not have the correct answers on the card OR he or she marked the answers in response to *the wrong questions,* then the game continues until there is a proper winner.

11. If you want to play again, reshuffle the cards and begin again.

Have fun!

TERMS/ANSWERS

=	22
+	24
7/9	26
5/8	27
5/7	28
1	30
2	40
3	44
4	45
5	50
6	70
7	80
8	90
9	95
10	100
11	ADDENDS
12	ADDITION
13	CARRY
14	DIGIT
15	HUNDREDS PLACE
16	ONES PLACE
17	SUM
18	TENS PLACE
19	WHOLE NUMBERS
20	ZERO

Additional Terms

Choose as many additional terms as you would like and write them in the
squares. Repeat each as desired.
Cut out the squares and randomly distribute them to the class.
Instruct the students to place their square on the center space of their card.

Addition Bingo

Clues for Additional Terms

Write three clues for each of your additional terms.

_____ 1. 2. 3.	_____ 1. 2. 3.
_____ 1. 2. 3.	_____ 1. 2. 3.
_____ 1. 2. 3.	_____ 1. 2. 3.

Addition Bingo

+ − X ÷	+ − X ÷	+ − X ÷	+ − X ÷	+ − X ÷
+ − X ÷	+ − X ÷	+ − X ÷	+ − X ÷	+ − X ÷
+ − X ÷	+ − X ÷	+ − X ÷	+ − X ÷	+ − X ÷
+ − X ÷	+ − X ÷	+ − X ÷	+ − X ÷	+ − X ÷
+ − X ÷	+ − X ÷	+ − X ÷	+ − X ÷	+ − X ÷
+ − X ÷	+ − X ÷	+ − X ÷	+ − X ÷	+ − X ÷
+ − X ÷	+ − X ÷	+ − X ÷	+ − X ÷	+ − X ÷

=	+
=	**+**
1. This sign means "equals."	1. This sign means plus.
2. Whatever is left of this sign has the same value as what is to the right of this sign.	2. Fill in the sign: 2 ___ 2 = 4.
3. Fill in the sign: 2 + 2 ___ 4.	3. This sign means that you should use addition.

7/9

1. $5/9 + 2/9 = $ ___
2. $6/9 + 1/9 = $ ___
3. $4/9 + 3/9 = $ ___

5/8

1. $1/8 + 4/8 = $ ___
2. $3/8 + 2/8 = $ ___
3. $5/8 + 0 = $ ___

5/7

1. $2/7 + 3/7 = $ ___
2. $1/7 + 4/7 = $ ___
3. $5/7 + 0 = $ ___

1

1. $0 + 1 = $ ___
2. $5 + $ ___ $ = 6$
3. ___ $ + 8 = 9$

2

1. $1 + 1 = $ ___
2. $5 + $ ___ $ = 7$
3. $11 + $ ___ $ = 13$

3

1. $2 + 1 = $ ___
2. $6 + $ ___ $ = 9$
3. $10 + $ ___ $ = 13$

4

1. $2 + 2 = $ ___
2. $3 + 1 = $ ___
3. $7 + $ ___ $ = 11$

5

1. $2 + 3 = $ ___
2. ___ $ + 7 = 12$
3. $9 + $ ___ $ = 14$

Addition Bingo

6 1. 2 + 4 = ___ 2. 3 + 3 = ___ 3. 6 + ___ = 12	**7** 1. 3 + 4 = ___ 2. 5 + 2 = ___ 3. 3 + ___ = 10
8 1. 4 + 4 = ___ 2. 5 + 3 = ___ 3. 6 + ___ = 14	**9** 1. 5 + 4 = ___ 2. 6 + 3 = ___ 3. 6 + ___ = 15
10 1. 5 + 5 = ___ 2. 6 + 4 = ___ 3. 10 + ___ = 20	**11** 1. 6+ 5 = ___ 2. 4 + 7 = ___ 3. 9 + 2 = ___
12 1. 7 + 5 = ___ 2. 3 +9 = ___ 3. 8 + 4 = ___	**13** 1. 7 + 6 = ___ 2. 4 +9 = ___ 3. 8 + 5 = ___
14 1. 8 + 6 = ___ 2. 5 +9 = ___ 3. 7 + 7 = ___	**15** 1. 11 + 4 = ___ 2. 6 + 9 = ___ 3. 8 + 7 = ___

Addition Bingo

16

1. 10 + 6 = ___
2. 7 + 9 = ___
3. 8 + 8 = ___

17

1. 12 + 5 = ___
2. 8 + 9 = ___
3. 11 + 6 = ___

18

1. 16 + 2 = ___
2. 9 + 9 = ___
3. 11 + 7 = ___

19

1. 10 + 9 = ___
2. 18 + 1 = ___
3. 11 + 8 = ___

20

1. 10 + 10 = ___
2. 18 + 2 = ___
3. 19 + 1 = ___

22

1. 11 + 11 = ___
2. 12 + 10 = ___
3. 20 + 2 = ___

24

1. 10 + 14 = ___
2. 12 + 12 = ___
3. 22 + 2 = ___

26

1. 12 + 14 = ___
2. 13 + 13 = ___
3. 24 + 2 = ___

27

1. 12 + 15 = ___
2. 13 + 14 = ___
3. 26 + 1 = ___

28

1. 26 + 2 = ___
2. 14 + 14 = ___
3. 20 + 8 = ___

Addition Bingo

30

1. 28 + 2 = ___
2. 15 + 15 = ___
3. 20 + 10 = ___

40

1. 38 + 2 = ___
2. 30 + 10 = ___
3. 20 + 20 = ___

44

1. 32 + 12 = ___
2. 21 + 23 = ___
3. 30 + 14 = ___

45

1. 31 + 14 = ___
2. 20 + 25 = ___
3. 40 + 5 = ___

50

1. 48 + 2 = ___
2. 45 + 5 = ___
3. 40 + 10 = ___

70

1. 68 + 2 = ___
2. 65 + 5 = ___
3. 60 + 10 = ___

80

1. 78 + 2 = ___
2. 75 + 5 = ___
3. 70 + 10 = ___

90

1. 88 + 2 = ___
2. 85 + 5 = ___
3. 80 + 10 = ___

95

1. 93 + 2 = ___
2. 85 + 10 = ___
3. 90 + 5 = ___

100

1. 98 + 2 = ___
2. 50 + 50 = ___
3. 90 + 10 = ___

Addition Bingo

Addends	**Addition**
1. The numbers being added in an addition problem.	1. The operation in which two or more numbers are united.
2. In the statement 5 + 2 = 7, the 5 and the 2 are these.	2. a basic operation of mathematics. The others are subtraction, multiplication and division.
3. In the statement 100 + 300 = 400, the 100 and the 300 are these.	3. Do this operation if you see a plus sign.
Carry	**Digit**
1. To transfer a number from one column of figures to the next.	1. Any of the numerals 1 to 9 and 0.
2. To add 34 + 7, you must __ one set of ten to the tens column.	2. 36 is a 2-___ number.
2. To solve 23 + 9, you must __ one set of ten to the tens column.	3. 268 is a 3-___ number.
Hundreds Place	**Ones Place**
1. The place three to the left of the decimal point.	1. The place just to the left of the decimal point.
2. In the number 927, the 9 is in this place.	2. In the number 17, the 7 is in this place.
3. In the number 534, the 5 is in this place.	3. In the number 215, the 5 is in this place.
Sum	**Tens Place**
1. The result of adding numbers together.	1. The place two to the left of the decimal point.
2. In the statement 5 + 2 = 7, the 7 is this.	2. In the number 248, the 4 is in this place.
3. In the statement 10 + 30 = 40, the 40 is this.	3. In the number 892, the 9 is in this place.
Whole Numbers	**Zero**
1. Positive numbers without a fraction or a decimal point are called ___.	1. It means "none" and is neither positive nor negative.
2. 5, 10, 32 and 447 are ____. -2, 10.5 and 3/4 are not____.	2. The sum of any number plus this is that number.
3. 7, 12, 312, and 547 are ___. -284, 14.5 and 0 are not ___.	3. 5 + ___ = 5

Addition Bingo

30	24	5/8	13	9
6	+	Zero	Addends	26
7/9	Hundreds Place		80	28
5/7	Carry	18	44	70
90	10	7	15	20

Addition Bingo: Card No. 1

Addition Bingo

5/7	7/9	40	Sum	Addition
70	Addends	2	Carry	45
100	10		8	18
14	17	Hundreds Place	16	26
20	Zero	7	6	15

Addition Bingo

5/7	18	Addends	44	7/9
10	+	3	24	22
Carry	Zero		45	=
Hundreds Place	100	90	14	40
15	6	7	16	Addition

Addition Bingo: Card No. 3

Addition Bingo

Hundreds Place	45	5/8	6	Addition
27	1	24	Sum	7/9
80	14		9	13
18	Digit	Zero	7	2
19	20	Whole Numbers	15	28

Addition Bingo: Card No. 4

Addition Bingo

20	9	Carry	2	6
27	18	3	8	+
5/8	28		50	11
26	Addition	30	16	19
Addends	7	7/9	Hundreds Place	80

Addition Bingo: Card No. 5

Addition Bingo

=	45	40	Addition	28
44	Carry	19	24	7/9
Sum	4		1	8
7	90	16	Whole Numbers	5/8
70	18	30	80	Digit

Addition Bingo: Card No. 6

Addition Bingo

30	45	11	50	Addends
70	Addition	10	+	27
40	13		8	1
Hundreds Place	14	3	5/7	100
7	6	16	Whole Numbers	=

Addition Bingo: Card No. 7

© Barbara M. Peller

Addition Bingo

80	45	5	44	1
27	5/8	Sum	28	2
Digit	95		Addition	9
15	Hundreds Place	5/7	19	14
Zero	7	Whole Numbers	Carry	70

Addition Bingo: Card No. 8

Addition Bingo

8	Addends	10	Digit	6
19	Addition	80	Carry	45
22	30		+	5
4	20	90	50	11
14	16	3	5/7	9

Addition Bingo

5/7	44	1	Sum	Digit
28	2	24	+	Addition
95	45		13	100
90	26	19	16	22
3	70	40	20	80

Addition Bingo

=	45	Carry	19	70
5	22	50	8	24
27	Addition		40	10
3	7/9	16	6	5/7
4	7	30	Whole Numbers	Addends

Addition Bingo: Card No. 11

Addition Bingo

Addends	9	22	44	8
10	Zero	5/8	Whole Numbers	+
30	11		28	Sum
7	14	Addition	5/7	27
45	5	95	4	2

Addition Bingo: Card No. 12

Addition Bingo

4	9	=	22	28
5/8	5	Addition	8	100
44	2		10	11
80	16	1	95	5/7
7	26	Whole Numbers	30	50

Addition Bingo: Card No.13

Addition Bingo

6	Addition	Carry	8	4
2	30	22	+	45
19	13		40	3
26	16	95	1	=
7	Sum	100	70	80

Addition Bingo: Card No. 14

Addition Bingo

50	8	Carry	Addends	44
=	40	24	5/8	19
28	30		7/9	45
7	22	5	16	4
70	14	Whole Numbers	Digit	10

Addition Bingo: Card No. 15

Addition Bingo

1	22	5	Digit	17
Sum	100	11	27	13
4	9		28	10
Hundreds Place	2	7	50	5/7
19	Tens Place	Whole Numbers	14	45

Addition Bingo: Card No. 16

Addition Bingo

3	Ones Place	12	22	6
50	19	16	13	11
8	80		Tens Place	5
20	70	5/7	Carry	100
90	4	Addends	44	9

Addition Bingo: Card No. 17

Addition Bingo

Digit	95	2	19	Sum
45	3	90	28	4
8	100		12	5/8
20	24	16	5/7	40
Tens Place	22	Carry	Ones Place	=

Addition Bingo: Card No. 18

Addition Bingo

28	=	22	5	95
50	44	45	Addends	13
Ones Place	6		+	7/9
40	Tens Place	90	14	12
5/8	17	70	80	Whole Numbers

Addition Bingo: Card No. 19

Addition Bingo

95	Ones Place	44	22	+
2	10	27	90	Sum
9	11		Hundreds Place	24
20	Zero	15	14	Tens Place
18	80	17	5/7	12

Addition Bingo: Card No. 20

Addition Bingo

50	=	27	22	26
9	12	1	5	30
100	70		Ones Place	Carry
90	Addends	Tens Place	20	80
Hundreds Place	17	Whole Numbers	3	14

Addition Bingo: Card No. 21

© Barbara M. Peller

Addition Bingo

Digit	40	12	5/8	4
Sum	44	7/9	5	+
2	13		30	11
Tens Place	20	14	24	6
17	3	Ones Place	100	27

Addition Bingo: Card No. 22

© Barbara M. Peller

Addition Bingo

1	Ones Place	Addends	5/8	Whole Numbers
=	95	70	50	24
40	4		15	30
100	17	Tens Place	3	14
26	Zero	80	90	12

Addition Bingo

1	95	6	Ones Place	5
12	Whole Numbers	27	Sum	30
11	Digit		4	100
26	15	Tens Place	3	9
18	Hundreds Place	17	44	Zero

Addition Bingo: Card No. 24

Addition Bingo

Hundreds Place	27	Ones Place	Carry	12
24	26	50	1	+
9	5		15	Tens Place
7/9	20	Zero	17	13
Whole Numbers	6	2	19	18

Addition Bingo: Card No. 25

Addition Bingo

12	Ones Place	40	Sum	Digit
90	44	5	95	1
26	15		13	Hundreds Place
3	5/8	20	17	Tens Place
11	19	Carry	Zero	18

Addition Bingo

40	2	Ones Place	95	10
26	15	50	Tens Place	+
16	Zero		17	Hundreds Place
Digit	=	27	18	24
4	13	12	7/9	11

Addition Bingo: Card No. 27

Addition Bingo

28	95	7/9	Ones Place	1
10	12	15	Sum	13
Zero	100		11	90
5/7	Digit	70	17	Tens Place
5/8	8	4	18	26

Addition Bingo: Card No. 28

Addition Bingo

12	95	Digit	50	8
26	90	27	11	7/9
9	15		+	Ones Place
10	20	Addition	17	Tens Place
1	5	18	=	Zero

Addition Bingo: Card No. 29

Addition Bingo

6	Ones Place	Sum	8	Tens Place
24	95	40	13	+
26	4		11	27
18	=	5/8	17	15
20	Addends	Zero	12	7/9

Addition Bingo: Card No. 30